] ....ing

HARNESS and SADDLERY

Major G. Tylden

Shire Publications Ltd

CONTENTS

INTRODUCTION

Twenty years ago it would have been difficult to interest the majority of people in this short study of the various appliances used by men through the ages in the employment of domesticated animals of many kinds. Of them all it is probable that the horse has always aroused the greatest interest, and it is partly because this interest has been so very much accentuated in recent years and also because of the great increase in tourism that this work has been undertaken. So many more people are today able to visit other countries and see for themselves horses and other types of animals at work and it is partly for them that this book has been written. One has only to go to Portugal or Spain to see oxen working as they have done for literally thousands of years and, although horse and mule harness has changed more than the ox yoke, it is still a thing of little change in a world full of little else. It is the same with saddlery; first principles remain—they took some considerable time to be fully understood—but men and women ride in so many different styles in various parts of the world that it is of interest to see how they do it and to know why.

Museums often help a great deal and so do statues; the difference between an equestrian statue of a king or a great leader of medieval times and one of a popular figure of the nineteenth century is very considerable, when one looks at the way he is sitting and the saddle he is using. Does anyone bother today about the occasional statuary group of a chariot? Apart from the historical interest the chariot was probably the first reason for men and women betting. For they bet just as hard in Greek and Roman days on chariot races as they do today on flat racing, steeplechasing and trotting. And the harness and saddlery of those days mattered just as much as do the saddle on a Derby winner or the complicated boots on a trotting champion.

State harness, used for ceremonial occasions, is still fairly common. Usually lined with velvet and decorated with medallions carrying crests and coats of arms, there are often sets to be seen in museums. There is a very fine exhibit in the museum on the island of Isola Bella in Lake Maggiore in Italy.

I have included a chapter on side-saddles, which would appear to interest many people now that they are so seldom seen—a notable occasion being Trooping the Colour at Horse Guards Parade when Her Majesty the Queen rides side-saddle. There is another chapter on bits and bridles, in many ways more intricate than they appear. In some countries they are embellished, notably mule bridles in Spain and Portugal, but in Great Britain they are severely plain.

A chapter on horseshoes has also been included because, with the spread of riding as a recreation, there is today a serious shortage of farriers, highly trained men, in this country. It follows that an increasingly large number of people probably take an interest in the subject of horseshoes, of which we know so little historically. People may also like to know what they are paying for when they have a horse or pony shod!

The illustrations are described separately. They have been chosen to show as far as possible in a limited space the main changes which have taken place through the years in the subjects dealt with.

Details of the saddlery and harness used by the armies of Great Britain and the Commonwealth are only mentioned in this book where this has been necessary to complete the overall picture. They are complicated and those people who are interested can study them in a recently published book, *Horses and Saddlery*, listed in the Bibliography at the end of this book.

THE TRACE, THE YOKE AND THE POLE

As soon as men and women had too many possessions to carry on their backs they inevitably looked for some form of traction. Presumably the dog came up for consideration. As a pack animal he is unsatisfactory, but attached to a rope, presumably of rawhide, with a sled made of the fork of a small tree several dogs could pull a considerable load, especially on frozen surfaces or sand. The dog will be mentioned again; he was, as stated, probably used on the trace instead of his masters and mistresses, as he still is. The trace is not really satisfactory for a single animal to pull with, but if a stick is put through the loop and a person or animal put at each end, considerable power is attained and the yoke has arrived. Yoke and trace are still with us. As far as we know the yoke was first attached to the horns of domesticated cattle with a trace in the middle between two oxen or bullocks. Usually there is a pad on the forehead and oxen still pull from the horns in Southern Europe today (plate 3).

The next step was to take the yoke back to the neck (plate 2), whence the ox exerts his power best, and to put short straight pieces of wood through it, one on each side of the neck, and join them underneath with a thong. We know from an old Irish myth that a king who was responsible for moving the yoke back from the horns was honoured with the nickname of 'Ploughman' because of the importance of the change! It may be noted that to exert braking power an ox lets the yoke move up against his horns.

The introduction of the plough introduced the pole, for it was simple to fasten the yoke to the long end of the forked branch, of which the short leg of the V was dug into the ground to cultivate it. The pole was of course the prolongation of the main beam of the plough, once the latter was shaped as it is today, with the pull from traces. So here must have been the beginnings of the trace or traces, the yoke for two animals as a general but not a universal rule, and the pole. The last named, used with all types of vehicles from sleds to waggons, was converted for a single animal by putting in two poles, the shafts.

By prolonging a pole or one shaft and bending the end upwards to take a thong running back to the horns quite a degree of braking power is attained. This can be seen in the

Trentino today. It is very necessary on steep tracks with wheeled transport.

It has been found that any number of pairs of yoked oxen up to eight, and sometimes more, can be used with advantage to shift heavy loads. All that is necessary is to strengthen and lengthen the trace to give room for more pairs of oxen, fastening it either to the sled or to the pole of the waggon or cart.

The wheelers, the pair nearest the load, are always the strongest and heaviest of a team and even without a pole can do a considerable amount of steering. The straight yoke gives more power than one would expect and the name has survived even in countries where oxen are no longer used. In England up to recent times it was usual to tell a carter to 'yoke' his horse, though there was no such implement to hand.

Once man could work iron, the plough became much more efficient and required more and more power to work it. Handles or a handle were added with which to steer it and the pole was in most cases no longer used.

The straight yoke, with side pieces running straight down as described, has remained the standard right through the centuries. Usually each ox had the side pieces joined under his dewlap by a thong of some sort, normally twisted before being adjusted. In some countries a long bar, like another yoke, went across from one ox to the other, connected up with the side pieces.

There was, however, an important innovation introduced, presumably at an early stage, which had repercussions on horse and mule harness which endure to this day. The yoke was slightly curved over the neck and, instead of the straight side pieces, others curving down round the neck more or less like a collar were fitted. Eventually these were joined to make a complete circle or bow, passed through the yoke like the old straight pieces, and secured above. These bows were used in England with the straight yoke just as fancy dictated. The ox, except in wet weather, does not gall easily and this new form of yoke, which could fit only roughly, seems not to have given any trouble. There is a good specimen of an ox bow in the museum at Rottingdean, near Brighton. The straight yoke was also in use in England in recent times. The first heavy artillery used in the field in modern times was pulled by spans of oxen and used by both sides in the Anglo-Boer War of 1899-1902.

HARNESS, THE HORSE AND THE CHARIOT

When first domesticated the horse, or rather pony, was not strong enough to carry a man on his back, but with intelligent breeding and feeding it soon became possible to drive him in a light wheeled vehicle. Known as a chariot, and in use from c.1700 B.C., by about 1500 B.C. it was drawn by two 13-hand ponies (a hand being four inches), had a pole about seven feet long, two wheels three feet in diameter with four or more spokes, and metal tyres with an axle four feet six inches long running midway under a light body with curved sides, holding at most three men, the charioteer (in Egypt he was usually of officer class), a bowman and possibly another fighting man. Used in large numbers the war chariot may be said to have been the forerunner of the armoured car. The psychological effect of a charge by large numbers of chariots combined the two great principles of warfare, fire power (from the archers) and movement. The chariot was only manoeuvrable on dry, level ground free of obstruction and was not suitable for mountainous regions. It was also used extensively for what might today be called big game hunting, for racing in arenas comparable to modern race tracks and, in Roman days, for travel.

It follows that the harness used had to be extremely strong, as simple as possible, and give the utmost power possible to allow the ponies to use the extreme speed of which they were capable. There was no question of 'discovering' harness; every charioteer and builder had known the ox yoke since childhood. To adapt it to the new and exciting form of transport was their job and so well did they carry it out that 'chariot rig' has survived in India, especially in Bombay, till today.

The harnessmaker, like so many capable craftsmen a man of markedly conservative outlook, took the straight ox yoke, lightened it, did away with the uprights altogether, gave it a length of about three feet six inches, bolted it to the pole with loose lashings as well and fastened it on the pony's or horse's withers—where the front of the saddle of today stops —on a small pad saddle. This saddle was girthed tight under the animal and a broad band went down the slope of the shoulder with a strap going from it, between the forelegs, to the girth. There were no traces; the pull on the pole was directly from the yoke.

The horse exerts his power of traction roughly from the

centre of the shoulder blade over which the 'broad band' going down the slope of the shoulder passed. Scientifically the solution of the chariot era was not ideal, but it worked, and worked well. We shall see how from this yoke harness better methods were developed, but the yoke rig undoubtedly gave very great possibilities of pace, for war and for sport.

The balance of a two-wheeled cart is always a difficulty as the pole must not be allowed to ride upwards. This the men in the chariot could control by shifting their position and it is known that in case of difficulty one of them would run along the pole to do whatever was necessary! Partly because the two pole ponies were vulnerable from the sides and partly to give increased power it was customary to span a pony either on one or both sides. This gave a team of four, the Roman quadriga, and at first these outriggers were attached to the chariot by one trace looped round the neck. Horses pull well from one trace even if not properly broken to harness. Sometimes a fourth fighting man was added to the crew of the chariot; he rode the outrigger horse on the right.

By Roman times the yoke was not only fastened to the pad but continued round the necks of the animals by curved pieces of wood—the adaptation of the ox-bow to the horse harness. The next step was to give each of the outside horses a girth and a pad with a breast band going from the girth horizontally round the shoulder and giving a straight and technically correct pull from the horse's shoulder to the vehicle by traces, one on each side. The traces were either hooked on to the vehicle or on to a short length of wood, known as a swingle tree, fastened to the front of the chassis, as it would be called today. These traces running from a wide breast band can be seen today in use all over the world by all types of animal used for traction. Breast harness had been developed very simply from the more complicated earlier neck yoke method.

Except in sport the chariot had had its day and was almost everywhere replaced by cavalry. They could operate in nearly all sorts of terrain and were much easier to train than charioteers. Also, as mounted bowmen, they were much more effective.

The first chapter dealt with ox harness, unchanged up to the present day, the second with the chariot, mainly used for war and sport, but as originally designed not really satisfactory for travel and much less for commercial use. Although for a very considerable time two and four wheeled vehicles had been in use, in the absence of roads they were probably better suited to the ox or the donkey than to the horse, pony or mule. Breast harness could be used for all these animals, although it was not practical in the case of oxen. Shafts are easily adaptable to the ox yoke, being merely attached to the ends of the yoke, and breast harness could be used with shafts by adding loops to a pad saddle placed further back (as in the case of a riding saddle) the loops holding the shafts in place as long as the balance was maintained.

The horse collar

Wheeled vehicles can carry very heavy loads but do definitely require some form of track or road. This the Romans provided to a limited extent between nodal points, protected by garrisons operating from central stations. The bulk of the growing commercial traffic of the world was carried on pack animals, which will be dealt with in a separate chapter, but on a Roman road, and probably on a few locally constructed tracks, wheeled traffic was possible. It was presumably soon after the introduction of roads that harness designers began to look to the ox bow that was used with the chariot and experimented with a close fitting variation of comparatively soft leather and other material. This was the so-called horse collar, very much in use today. It is made of leather stuffed with straw, shaped to fit exactly to the neck of the animal and is thick enough to allow 'harness' bars of steel or wrought iron, with rings to take traces, being buckled round on the outside. This gives a most correct form of harness and allows an animal to develop its power of draught to the fullest extent. It is very suitable for single draught whether in shafts or with pole harness and it can be fitted to most types of animal, including oxen. Collars were in use on oxen in Great Britain early in the last century.

A horse or mule, the heavier the better, with collar harness in single draught, can pull very considerable weights indeed on a well metalled road and those people who have been fortunate enough to see heavy timber being hauled out of a wood will have realised how much strength a horse can put into his work when using a properly fitted collar.

The horse collar has been acclaimed by modern authorities as one of the greatest discoveries ever made. This may be so, but the balance of proof by experience makes the statement very doubtful. The collar has never to any great extent ousted the simpler and less efficient breast harness and presumably never could. The wide band round the neck low down on the shoulder of breast harness can be adjusted from one size of animal to another in a few minutes by raising or lowering the strap which goes up over the neck at its base. The flat surface of the breast strap fits any shoulder. Any worker in leather can make a band of this sort without any special training.

The case of the collar-maker is very different. He must be and always has been a carefully trained specialist. No one animal's neck where the collar has to fit is the same as any other and though collars can be changed and made to work on more than one animal, this is not usual. The collar-maker fits or alters the collar for each animal and if he knows his work there will be no undue friction and no sore shoulders.

It is because of this necessity for altering collars that most of the horse-drawn artilleries of the world gave up using them. Losses in action or from overwork might make it impossible to transfer collars to fresh animals, though with breast harness this gave no problems at all. As we shall see, in coaching days the horse and his collar belonged together.

Breeching

So much for the problem of draught, solved at some unknown date, though doubtless by a very gradual process from area to area. We now come to the necessity for the animal or animals being able to stop a vehicle if necessary.

As long as the yoke was in use with horse and mule harness it would be used to exert backward pressure, but only in exceptional cases, to be discussed later, was the yoke retained with either breast or collar harness. In both types a strong strap, or in some cases a chain, ran from the centre of the breast band or collar to rings on the end of the pole. Running back either from breast band or collar was a broad band going round behind the animal well under the tail. In most cases this breeching, as it was called, had a strap or straps behind the pad saddle, going over the back. When the animal pulled back from the pole and 'sat', as the expression went, on the breeching under the tail, it is obvious that a considerable amount of force was exerted towards retarding the vehicle. In single harness with shafts the leather loops through which the shafts passed and which were on the pad saddle were connected with a short piece of breeching which was fastened on to the shafts by a short strap or chain.

TOWARDS BETTER ROADS: FROM THE MIDDLE AGES TO THE EIGHTEENTH CENTURY

The Bayeux Tapestry shows collar harness in use in agriculture by both horse and mule, but only in draught from swingle trees. There was, however, another type of harness used in single draught with shafts, and it has survived to the present day. Known as farm harness and used with the horse collar it has also been a great deal used in heavy

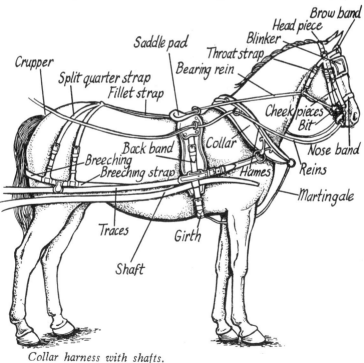

Collar harness with shafts.

draught by brewers and hauliers. The pull is from a short chain on the collar hames hooked on to another short chain fixed on the shaft, which is supported by a chain running over the pad saddle. A short length of breeching, also held up by the pad, is hooked on to the shaft to give braking. There are no traces. There are illustrations of this rig in Britain in the fifteenth century. One shows no pad saddle and it is very probable that it was in use much earlier. It was usual in all types of harness to run a long strap from behind the pad saddle divided in half so as to pass under the tail. Two vertical straps held up the breeching and the piece round the tail was known as the crupper. This arrangement presumably came in with the introduction of breeching. With heavy vehicles it was usual when negotiating a very steep hill to take out the team in front of the wheelers and hook them on behind to steady the descent.

The coach

The coach, a large weighty four-wheeled vehicle with the body slung between ropes or strong leather thongs, introduced a new era into the history of harness making. A coach had to travel at speed, and both horses, normally four with more added on very bad stretches, and harness had to be of as good quality as possible. Teams were changed frequently and it was far more satisfactory to provide harness, if possible, for each horse, especially if the collar was in use, as it certainly should have been for the wheelers. The coachman became a man of importance, highly skilled, who would look over the harness with a professional eye. Hooking on an extra pair of horses for a heavy stage usually meant that the near or left side horse would be ridden by a man of light weight, known later on as a postilion, with his right boot reinforced by a bar of metal or wood to prevent the harness of the off or right-hand horse rubbing or damaging his leg. Both coachman and postilion had a very practical knowledge of saddlery and could replace a broken trace in a very short time.

Artillery

There can be little doubt that the growth in importance of artillery, which coincided with the introduction of coaches, gave an added importance to transport generally. In Germany in the latter part of the fifteenth century, guns were mainly transported on heavy waggons, on which they were loaded by means of cranes, drawn by teams of three, four or five pairs of horses, harnessed to a pole, usually with a long chain trace, some with postilions and all usually shown in breast harness. Wheelers often had collars, and later all the horses would be fitted with them, with traces running back to swingle trees on the breasts of the pair behind. Light French guns had shafts fixed to the trail of the gun by 1525. The trail is the extension of the gun carriage which supports it and was later hooked on to the small wheeled cart, the limber, from which the wheelers pulled. With shafts on the trail only one horse could be used and the weight on its back was very severe. It was a long time before trains of artillery, as they were called, came to be horsed in peace time with their own horses, harness and properly enlisted drivers. When war broke out horses and drivers were commandeered and had to take their place among the soldiers, who could not carry on without civilian help. So no study of civilian harness can be complete without frequent references to artillery! It is not to be wondered at that, if the civilian carters and their teams came under fire which they considered severe, they cleared off with their horses and harness and lived to fight another day!

In England in Stuart times there were far more vehicles on the roads; both long waggons and coaches were increasing, but there can have been little or no changes in harness till the end of the eighteenth century.

MODERN ROADS : THE NINETEENTH CENTURY AND AFTER

The first quarter of the nineteenth century saw the reappearance, in a better form, of the Roman roads. Engineered by two Lowland Scots, Telford and McAdam, they transformed the whole of the industrial and commercial life of Great Britain. The better the roads were, the faster the speeds possible and the more numerous the vehicles! So the saddler and harness maker, to say nothing of the farrier, became of more and more importance. Stronger, better looking and more efficient harness was the keynote and in what may fairly be called the coaching age, a large section of the community was very much concerned with harness. To handle with skill and distinction the ribbons, the long reins of the coach and four, was the ambition not only of the professional coachmen but of anyone who could afford time to learn the art, for an art it was considered. From peers to young squires and sons of well-to-do farmers, a knowledge of harness was considered very much the correct thing. Books were written on the subject and even the advent of the railways took a considerable time to make much impression on the importance of the world of carts, carriages and harness.

Mail coaches and the expensive post-chaises, could travel up to twelve miles an hour over the scientifically graded highways with culverts, bridges and posting stations. The less weight a coach horse carried, and still carries, the better. He had to trot fast and in the old days there were what were known as galloping stages, where the coachman 'let 'em out', especially if he was tooling 'three blind 'uns and a bolter'! For although such a team was not common, it was not unknown! Coach horses (see plate 9) wore blinkers on their bridles and collars with pole-chains (sometimes with spring-hooks instead of ordinary hooks), although private coaches sometimes had pole-straps—leather traces with a chain-trace carried on the coach in case one broke. Pad saddles were used on later coaches, but an old line drawing of c.1880 shows none on the leaders, and in some cases a martingale runs from the collar between the forelegs to the girth of the pad saddle. Some authors advise chain-traces and list them. The wheelers' traces went to a draw bar on the body of the coach, and those of the leaders to swingle trees on each end of a longer bar hooked on the pole. There are no breechings, as the coach has a powerful brake. Coach harness was normally black with brass fittings, brown leather being used for the reins.

State harness, which is sometimes to be seen in museums, is much more complicated than coach harness. Everything is of leather, often lined with coloured cloth or some other material. There are always breechings and a double girth to the sometimes very ornate pad saddles, and sometimes breast harness instead of collars.

So often in the long history of harness, the requirements of war tended to increase efficiency, but in the case of the mail coach it was the other way round; the vastly improved roads brought rapid, controlled movement of vehicles and from the methods of the big English coaching companies, those

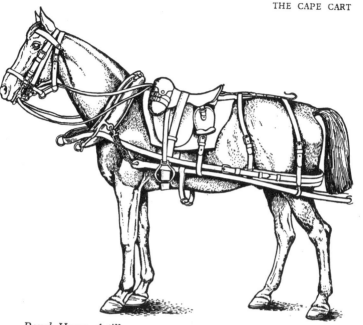

Royal Horse Artillery gun team near wheeler, 1898. (Gilbey, 'The Harness Horse', Vinton, 1898).

responsible for horsing the artillery learnt how the guns should be permanently staffed with the right sort of 'cattle'—in the slang of the day—with harness and with trained men to drive the teams. We have with us today in London the ultimate expression of moving field guns at almost racing pace in the King's Troop, Royal Horse Artillery.

The Cape Cart

Up to this time it had been considered obligatory to have shafts for at least one pair of wheelers with a two-wheeled vehicle in order to prevent it tipping. This was the case in the Royal Artillery, until they went over to pole draught at the end of the nineteenth century.

Meanwhile, at some unspecified date, but fairly early in the eighteenth century, some extremely knowledgeable but, alas, unknown cart and waggon builder in South Africa, in what was then the Cape of Good Hope, had evolved what is known as the Cape Cart rig (see plate 8). It is still very

15

much in evidence. A two-wheeled cart, taking either two people—a buggy—or four or even six—a Cape Cart, has normal breast harness with traces and straps to the end of the pole. The balance is maintained by running a light yoke of some very strong wood either under or over the pole with straps round the pole which allow the yoke to have play. A strap goes from each end of the yoke over each horse's neck, each strap fitting round the yoke close to the breast band. This very easy and most efficient device was in due course used by the artillery with their two wheeled limbers and can be seen on the King's Troop, Royal Horse Artillery.

The curricle

The Cape Cart itself never caught on in England. There was in fact little use for a light two-wheeled cart with a pair of horses. Among the multiplicity of all sorts of vehicles, introduced after 1800, landaus, barouches, phaetons, tilburies, buggies and gigs, there was only one two-wheeled vehicle for a pair. Known as the curricle and originating in Italy, it was in high favour with the dandies of the Regency period. Readers of Georgette Heyer's romances will readily recall many affectionate references to this fashionable vehicle. The curricle was fairly high, as it was intended for a pair of very fast, biggish horses. It took two people, with often a seat behind for a groom. The harness was normal with collar, traces and pole straps, and a strong pad saddle. Across from one saddle to the other ran a light metal bar, often silver plated. The bar went through a metal ring on each saddle and was connected to the pole by a strong spring with straps. This arrangement supported the pole and was not intended to have anything to do with moving the cart. As a matter of fact, if badly adjusted with traces not of the proper length, some of the strain of pulling might come on the bar, which was very unsatisfactory. It was not considered a suitable rig for bad, dirt roads, but it was very smart and a pair of horses in a light cart of this type could really travel! Specimens survived till about 1880 and an occasional one could be met with as late as 1900. There is one specimen in use today at Goring, Oxfordshire, where a firm specialises in old vehicles with their harness. There are still quite a number of horse-drawn carts and carriages to be seen at shows and on the roads at certain times, although generally speaking they are now museum pieces.

Trotting races

In various countries, especially Russia and the United

States of America, trotting races, with horses harnessed to the lightest possible sulkies, were immensely popular right throughout the nineteenth century and still are. There are also, to a lesser extent, pacing races, with the horse moving alternately the right and left, or left and right pairs of legs. The light two-wheeled sulkies need the least and lightest amount of harness. Breast harness with traces or simply a very small pad saddle with the traces running from it is normal, there is no breeching and the most remarkable parts of the tackle are the various rather complicated forms of boots worn on all four legs to prevent the horse, when trotting a mile in under two minutes, from 'interfering' with himself, as it is called when one leg strikes another, causing serious injury. The late Captain Maurice Hayes has this to say about the problem: "The drawback of 'hitting' is greatly obviated by the use of 'boots'; in fact 'boots' make the trotter, as the Americans say. Almost all fast trotters have to wear several kinds of boots when doing their best; but at a slower speed, they do not require them as a rule. Many become so well aware of the danger of hitting that they will not extend themselves if their legs are not thus protected. 'Grabbing' boots are used to protect the heels of the foreleg; 'scalpers', the toes and coronets of the hind legs; 'quarter' boots, the coronets in cases of brushing; and 'ankle' boots the fetlocks. 'Knee' and 'arm' boots are employed against speedy cutting; and 'elbow' boots against the elbows being hit, in which case the animal is liable to get both elbows capped. Knee, arm and elbow boots require 'suspenders' to keep them in place".

Deer and dog harness

There are two more types of harness to be considered—those used for reindeer and for dogs. In Lapland the half-wild deer are driven in sleds with a leather harness saddle from which, about halfway down the animal's side, a wide strap which serves as a collar runs forward to the front of the small hump on the neck. It carries a rawhide trace plaited square. There is a quick release on the sled and everything is as simple as possible.

Although it is illegal to work dogs in harness in Great Britain, this practice is normal in various countries, some in Europe. Plate 4 shows a Swiss mountain dog pulling a light cart loaded with milk churns. He has a wide collar with tugs on it to hold the shafts and rings for wide leather traces, which are held up by back straps on a light luggage saddle.

The tackle used in North America with the 'huskies' that pull sleds has either breast harness or collars. The latter are 'carefully fitted'. The traces are sewn to a belly band going on a back strap. Each dog can pull between 100 and 200 pounds, according to the state of the trail.

This concludes the account of harnesses. As should be apparent, the changes over many centuries have not been as many as might be expected. The trace, the yoke and the sled remain and, as a reminder, the bays so familiar in the facade of Elizabethan houses still show the natural curve of an ox drawn sled with no pole swinging in as close as possible to deliver stone to the builders. Driving a sled is not that easy!

A satisfactory point about harness is that both from watching it in use and from studying it in museums, it is usually obvious how it works. One can see how the wheelers of a pair pull back from the pole and how both in single and double harness, a horse literally sits on the breeching to stop the vehicle!

EARLY RIDING SADDLES

To very many people a saddle, the man-made contraption which has allowed millions of people to travel, to fight, and to take their pleasure in comfort on the backs of horses and to experience the delightful sensation of speed, is one of the most useful and romantic inventions ever made.

The difficulty in discovering saddles is to realise just how they work and unless one sees a saddle taken to pieces it is not easy to understand. In nine cases out of ten the body of the rider covers the details and one has only a general impression. A well-made saddle should fulfil two functions. First it should fit the horse in such a way that in none of the animal's complicated movements should it cause pain or discomfort, least of all injury, and it should not be heavier than is absolutely necessary. Secondly it must allow the rider to indulge in whatever roles he may wish, with competence and in comfort. For his own ends man has also often insisted on fastening a number of, to him, important objects on to the saddle. (It is well to realise that the less weight put on a saddle, the better it is for the horse!)

Once the first function is carried out and the saddle fits the horse to perfection, the second one, the comfort and efficiency of the rider, can be attained in a number of ways and the rider can sit in many different positions. There is no absolutely perfect, fool-proof way of riding, although very large numbers of people will not agree. There are many and various styles of riding and they all can and do produce undeniable horsemen and have always done so!

Pad saddles

Up to about the fourth century A.D. saddles were simply pads made of some durable, comparatively soft material secured by bands passing under the horse—the girths; round the chest—the breastplate; and round the quarters under the tail—the breeching. On this the rider sat in the natural position as shown in the sculptures known as the Parthenon Frieze. The pad was not more secure than riding bareback, but it was more comfortable. It was with the pad that Alexander the Great's cavalry ousted the chariot from its place in the line of battle and put the mounted arm well on its way to a pre-eminence it was to enjoy for many centuries.

Three types of pad saddle come up for examination; one nearly two thousand years old, the other two in use today. In the Hermitage Museum, Leningrad, is a cast gold relief showing Scythian horses saddled up with their riders seated or lying down holding them. The saddles have only rudimentary arches and these, like the seats, look to be ribbed with some substance which gives the effect of quilting. Girths go over the saddle like surcingles and there are curiously shaped breastplates with straps hanging down from the cantles like the ties on a cowboy saddle.

The next pad saddle is the Argentine *recardo* or riding saddle of today. Folded blankets are placed under two sheets of leather with rounded ends. Two straw pillows eighteen inches long, joined by thongs, go on each side of the backbone. A broad rawhide girth goes over them with rings for stirrups. On top goes a sheepskin with a surcingle.

The last is a donkey saddle as used today in the Near East. Flat at the cantle, which goes just in front of the tail, it is padded up high in front with a central strip of leather running across the seat front to back. There is a girth, but no stirrups and the rider sits, correctly, right back as shown in plate 1.

The saddle tree

We now come to the saddle tree, thought to have been evolved late in the fourth century A.D. The Mongols claim to have invented it and it is understood that the tree they use today has remained unchanged. It is both simple and effective as well as revolutionary. Originally made of wood and rawhide it has since been fashioned from many materials, especially steel, and today springs are incorporated. The framework, for such it is, must fit the horse's back behind the withers and rest on the ribs.

A saddle tree consists of four parts; the two side-bars, flat pieces of wood shaped to the ribs on which they rest, are supported front and back by two arches, which originally ran from the bottom of the side-bars to varying heights over the spine. A horse's spine is only covered with skin and is very vulnerable to abrasions—the object of the framework of the tree was to give the rider solid support but to keep some of his weight off the horse's backbone. In time all weight was kept off it. Various forms of padding were added to the tree to prevent rubbing, either in the shape of folded blankets or padding sewn into suitable material and fixed to the tree. The front arch was known as the pommel or the cantle, alternatively in England as the *arçon,* the word used for both arches in France. The rear arch was called the cantle and could be made in varying shapes—round, curved, stretching down on to the side-bars or rising sharply to a peak or 'spoon'. The front arch or pommel was in time developed into the horn of the cowboy's saddle. Both arches were often made high enough to protect the rider's vitals from enemy action. In time the side-bars were prolonged in front of the pommel to take a load. This extension is called the burs. When extended behind into what are called fans, they not only take loads such as rolled cloaks etc., but are of great use in keeping the cantle from pressing down on the backbone.

We now come to the second most important innovation made to the tree. This is the continuation of the arches below the bottom of the side-bars into what are called the points. These give a very considerable degree of stability to the saddle and are of considerable importance. In England from the time of the disuse of armour, only the front arch had points; on the Continent practically all saddles had points, back and front, well into the eighteenth century. Some makes of saddle never had points. Today in some types long stout pieces of felt incorporated in the stuffing of the saddle take the place of points to the front arch. As time went on the tree was so constructed that there was a gap, the chamber, right along the backbone, so that the whole weight of the rider rested on the leather seat of the saddle, a very great improvement which became general in some types late in the eighteenth century. Before this, too much of the rider's weight rested on the backbone.

Stirrups

While the saddle tree was still in its simplest form late in the sixth century—earlier dates have been put forward—the most remarkable improvement to the saddle was introduced by the invention of stirrups, loops of strong material shaped to fit each foot, hung from the saddle tree either from the centre of the side-bars or further forward. The further forward the stirrups hang the more bent the rider's knee can be. With the stirrups in the centre position the leg is straight and the rider is virtually standing over the horse. The great support given by the stirrups made it possible for the horseman to use lance, sword, mace, battle-axe or polo stick with a strength denied him before. This was especially the case with the lance. With the feet firmly supported it could now be used to advantage, especially when couched—as the expression is—well under the arm. It could then be driven home with the full force of horse and rider moving at speed. Nothing, not even plate armour, could take the shock successfully. In the Bayeux Tapestry some of the Norman knights do not seem to have attained this skill and are shown using their lances as throwing spears. The stirrup gave cavalry a pre-eminence over men on foot which lasted well after the invention of gunpowder, in fact into the nineteenth century. The sword, especially the curved sabre so liked by the peoples of the East, could now be used with greater force, for a man riding with short stirrups can stand right up in them and deliver his blow with deadly effect as he reseats himself in his saddle. Today it is hardly necessary to remind people of the great strain a modern flat race jockey puts on his stirrup leathers.

As far as we know, the saddle used with chain mail had a rectangular wooden tree with no points to the arches, often with two girths and always a breast band. The arches had curved tops and the seat was the straight-legged one made more or less obligatory when armour, especially any sort of leg armour, was worn.

The Mongol saddle

The incursions of the various nomadic tribes from Asia, especially those of the north-eastern zones, brought the heavily armoured European horseman sharply in conflict with people using a completely different method of riding. From furthest Asia to the southern side of the Mediterranean the horseman liked a saddle with not much room in the seat, with stirrups set forward for a well-bent leg, arches of medium height, no points to the side-bars and often a considerable thickness of saddle cloths over the seat. In some cases the saddle was well and permanently padded, in others a thick blanket separate to the saddle was customary. The Mongols claim to have invented the saddle, which as far as is known seems likely enough. It was essentially suited to the all-round firing positions of the mounted archer, and it served equally well for the swordsmen who drove home on the enemy when the archers had softened them up. Like the first Duke of Wellington, who said that at the age of seventy he could 'make the saddle on his horse's back his house', the Mongols liked to live in theirs and did so, and early in the thirteenth century defeated the mailed might of Central Europe in one battle after another.

The side-bars of the Mongol saddle are straight with rounded ends and are deep with no points and short burs and fans. The arches are high and the pommel gives a considerable amount of protection. With a short stirrup it gives an almost crouching seat with the feet drawn well back, so characteristic of Eastern peoples. They ride incredible distances and expend great ingenuity and often a high degree of craftsmanship on their saddlery, especially the stirrups. These are often what is called the slipper shape, made hooded with a flat foothold taking half the foot only. A peculiarly horrible Eastern type had the inside of the all metal stirrup made with a sharp edge, which would cut the horse's flank open if the slightest pressure was applied.

None of the saddles already described were particularly heavy. They were strong, simple and easy to make. But the change-over from chain mail to plate armour which began c.1300, made stronger and heavier saddles inevitable and by the last quarter of the sixteenth century, when plate armour was going out, they were very heavy indeed. The saddler, working with the armourer, had to cope with a total weight on the horse of, in the case of tournament armour, as much as four hundredweight, including the weight of the horseman. The saddle could weigh as much as sixty pounds and even with half armour a very strong tree was necessary. A man-at-arms in plate as worn in the field in a campaign need not have had a greater weight on the horse than twenty-two stone, though the saddle would have not been less than forty pounds.

Horse armour

We are fortunate that in the Tower of London and the Neue Hofburg in Vienna we can study plate armour in such profusion, and the *Inventory and Survey of the Armouries of the Tower* gives us descriptions and reproductions of photographs showing every detail of 'saddles and saddle-steels'. To carry the weight the saddle trees had to be very strong and large and had two, three and four plates to each arch, mostly made of steel and decorated with designs to match those on the armour worn by the man, as was the horse armour. Sometimes the plates were made of bone and in other cases the tree itself was made of ivory to which the plates were fastened. The horse armour was designated as follows: the *chanfron*, covering the front of the head below the ears; the *crinet*, laminated plates covering the mane right down to the front arch; the *peytral*, three deep curved plates running round the chest with a glancing knob to deflect lance blows; the *crupper*, two deep plates protecting the whole body behind the saddle (from this is derived the modern term 'crupper'); a strap running from the saddle round the tail called the *tail guard;* the *flanchards*, plates hanging down below the saddle covering the girth, and the *rein-guards*.

Today the Household Cavalry still have the top plate of the crinet in the form of metal scales attached to the top of the bridle to prevent it being slashed off by an opponent and up to quite recent times British cavalry had chains to fasten the horses to the manger by the halter. On active service these chains were fastened on one rein to stop it being cut through. Both forms of rendering a cavalryman's horse unmanageable were frequently employed, especially by the Afghans in 1879 and 1880.

Interesting though these wonderful sets of horse armour for the jousting are, they would have been impossible to use in warfare. An outstanding representation of a medieval war saddle is to be found in Dürer's watercolour of the *Gewappneter Reiter* of 1498 (see plate 7). The rider is in three-quarter plate armour and his saddle is drawn in most carefully. The first things to notice are the points back and front to the very sensibly-sized pommel and the cantle which obviously gives a fair measure of support to the heavily burdened horseman. The stirrup leather, another favourite target for a swordsman, especially a dismounted one, hardly shows below the flanchard. Note that both breastplate and crupper are shown and also a loose breeching strap. This may have been a fitting for the so-called textile trapper, sometimes used instead of horse armour.

It is thought that in the thirteenth century trappers of chain mail may have been used, although how a horse managed to walk, let alone trot or gallop, hampered in this manner, is not clear! The textile trapper or covering for the horse was very frequently used and sometimes churches were despoiled of their embroideries to provide them. Breastplates of leather were also used and in the seventeenth century, when armour was going out of use, it was common to have a *pattrell*, a narrow curved steel plate on the breastplate, the last vestige of the *peytral*. The pattrell would stop a lance thrust or a sword cut.

The Duke of Newcastle, c.1660, showing the special saddle used for the manège and the straight-legged seat. (S. Sidney, 'Book of the Horse', 1878).

RIDING SADDLES : THE SEVENTEENTH CENTURY AND AFTER

We now come to a period when quite a considerable amount is known of the saddles used by civilians. Although there are no dates available, there is little doubt that the so-called English saddle, the hunting saddle of today, now being to some extent replaced by a newer type but still very much in evidence, was in existence in the reign of Queen Elizabeth I. It is not known when it was first in use but in its long life it must have been made and used all over the world in tremendous numbers. Almost flat, with a slightly raised pommel, this saddle has the stirrups set some distance in front of the centre of the side-bars. The weight is between ten and fourteen pounds, the channel along the spine is small and to sit in the saddle gives the impression of being close to the horse, as one undoubtedly is. From 1600 onwards there are quite frequent mentions of scotch saddles, which were also popular all over England. It is known that they had very broad flaps, and that the padded panels were larger than those of the English saddle. Also in use were Hackney saddles with iron trees, and pad pannells presumably without trees.

The 'Great Saddle'

The last knight-in-armour type of saddle was still in use for manège or riding-school work. Very narrow in the seat, it was described by the Duke of Newcastle in 1658 as 'so well made that a man must sit upon it with a good grace whether he will or no'. The saddle weighed about sixty pounds and it was padded 'like the body of a well padded library chair' (see illustration). Not a very practical saddle for everyday use, it may be seen in a vastly attenuated and sensible form in the Spanish Riding School in Vienna. As will appear, it passed through many vicissitudes before reaching its present form. Known as 'the Great Saddle' it was in very limited use in the Royalist cavalry during the Civil Wars, when the military-type saddle most in use was of Continental origin. By the beginning of the seventeenth century cavalry was wearing half armour and using heavy pistols from the saddle. The saddle which evolved was a very much lighter affair made of leather, with a flat seat and low arches, mostly with short points. There were padded rolls supporting both thighs and knees and the amount and size of these aids varied in different countries. There were always breastplates and breeching and the military models had hooks on short burs to take the long, heavy pistol holsters, one on each side. Usually the holster passed through a loop on the breastplate to hold it steady. In use, in some cases as late as the last quarter of the eighteenth century, this saddle was adapted for civilian use with only small rolls and with the points of the arches coming down outside the seat instead of underneath it. Sometimes the seat was quilted and the general effect was of a rectangular shape. An exhibit of the late seventeenth century at Edinburgh Castle shows considerable padding. It was earlier known in England as Markham's saddle. An ex-officer with Continental experience, Markham was also a prolific writer on military subjects.

By the mid eighteenth century civilians were using pistols instead of swords as a means of personal protection. Normally made in pairs with a belt between them going over the pommel, pistol holsters had covers, usually of fur. They had to be substantially made because the so-called 'horse pistol', originally carried by regiments of horse or heavy cavalry, was of large bore with a long barrel. Prior to this the man-at-arms carried a mace or battle-axe at his pommel and in the seventeenth century a riding cloak was carried rolled on the cantle. The Romans had used saddle bags, and the most usual type, still in use, was double bags hanging behind each leg with a wide band over the seat.

The Hussar Saddle

With one exception, the trend in all European saddlery during the eighteenth century was away from the so-called 'high mounting' knight's saddle with its very high arches. In England the newly introduced light cavalry had adopted a rather stronger version of the English hunting saddle, also well known in France. The exception was the introduction by Hungarians, who had been serving in the Austrian cavalry as border guards since 1526, and were known as Hussars, of the deep-seated cavalry saddle with high arches which is still in use by the Household Cavalry and the King's Troop, Royal Horse Artillery. Known for many years as Hussar saddles and first used in the British army in 1805, these simple, strongly-built saddles were almost certainly derived from those used by the Mongol armies which flooded into Hungary in the thirteenth century A.D. The similarity between the tree of this saddle and that of the Mongol type, still in use in Mongolia (see page 21), is very great. The tree, as introduced in the British army, can be studied in detail in *Horses and Saddlery* (see Bibliography).

The model or hunting seat of 1870 and later also known as the 'squire's seat' (S. Sidney, 'Book of the Horse', 1878).

As adapted by the Hungarians the tree became standardised and introduced a great improvement into all saddlery. Prior to this a great deal of the weight of the rider bore directly on the horse's spine, although the side-bars took a proportion. By lacing the gap between the side-bars with rawhide stretched tightly, as is the webbing which in time superseded it, the

27

channel or gullet along the spine was left completely free of any of the weight of the horseman. The side-bars were well stuffed or padded. The deep dip in the seat and the high arches allowed the rider to sit well into the saddle and to ride with a fairly short stirrup. Leather flaps were fixed under the stirrups and a leather seat, at first loose and quilted, later also stretched, completed a type used by most cavalries right up to the present time, and found in use in some of the many small riding-school establishments of today. Its use gives a rather stiff version of the old English hunting seat.

A distinguishing feature of the Hussar saddle is the high, narrow 'spoon', rising from the centre of the cantle and allowing much too much 'luggage' to be carried! Until comparatively modern times there were no points to the arches. They are now replaced by points made of thick felt and the tree is made of steel. There have always been burs and fans and the amount of luggage carried has been a very serious burden to the unfortunate troop horses. This saddle not only spread over Europe but was adopted in America. Its great point was the way the seat was raised off the horse's backbone. It was usually much disliked by hard riding, hunting and steeplechasing men, who detest being unable to 'sit down in their saddles and ride like hell' as the pigstickers in India used to put it. The hunting saddle was and is their ideal.

The colonial saddle

A very important modification of the hunting saddle is the so-called colonial, staff officer's or yeomanry saddle. Its trade name is the Cape fan, although who first made it, presumably at the Cape of Good Hope, is unknown. It was in use in many parts of South Africa by 1877 and was officially adopted by the British army in 1879. It is in use all over the world, especially by mounted police, including the Metropolitan Mounted Police. It is simply a hunting saddle with the tree prolonged into burs and in some cases rather pronounced fans. Normally it has padded panels which are strapped or screwed in place and can be removed in a few seconds for restuffing. Although quite an amount of luggage can be carried on the colonial saddle, it was not designed for nor should it be used with the excessive weights carried by regular cavalry. It is a nice looking saddle and is to be found, sometimes with much shorter fans, in many parts of the world. While the Hussar type can be and often has been used as a packsaddle (see page 42), the colonial saddle cannot.

Like the hunting saddle the colonial pattern has the stirrup leathers attached to the side-bars just behind the points. They fit round a small metal frame with one end closed by a spring-operated catch. This catch should never be kept shut when there is a rider in the saddle. If closed the catch can very easily jam and in the event of a fall the stirrup may not come away from the saddle as it should, jamming the rider's foot in the stirrup and causing an accident. There is no release catch fitted on cavalry type saddles.

In Australia and New Zealand a special design known as a buckjumping saddle has been in use for many years. It is shaped more or less like a hunting saddle, has a peculiarly deep dip in the seat (the dip is measured by dropping a perpendicular from a line drawn from pommel to cantle), a narrow waist (the narrowest part of the seat) and very large knee rolls curved back so as to fit against the thigh above the knee. There are also large rolls curving down from the cantle behind the thighs.

American saddles

The medieval knight's saddle has continued in varying forms in both North and South America. The first horses brought to the continent were the mounts of the Conquistadores, Spanish men-at-arms who brought their own heavy saddles with them. This was early in the sixteenth century and it is from these that were developed the stock saddles used by nearly all men who work with horses in the Americas. The very high cantle, the deep seat with the stirrups set well in the centre for a straight legged seat and the wide side-bars continued behind the saddle are typical. But the pommel has been developed into a very thick horn with a flat top designed to take the lasso, the rawhide 'rope' with which the horse or bovine is caught in the open. The wooden stirrups are large and open and the weight of the saddle is considerable, fifty pounds and over. These saddles are tremendously strong, as they must be, and men who are used to them are said never to take kindly to other types. They do, however, vary slightly in different localities.

The United States cavalry troop saddle, which derives from the Hussar saddle, bears no relation at all to the stock saddle, except that neither of them have points to the arches. It consists of a plain tree, rather narrow, with very short burs and fans, fairly high arches and only a slight dip. The centre between the side-bars is open and the girths, known as cinches, run from the burs and fans to a central ring on which the cinch itself is laced.

The 'forward seat'

The great revolution at the end of the nineteenth century

was the so-called 'forward seat', first used in flat racing and introduced into Great Britain from the United States by Tod Sloan, a professional flat race jockey. Although he was not the first man to use it in England, he made it famous and it came into general use in racing. It was an Italian cavalry officer, Captain Caprilli, who introduced this crouching seat and another Italian officer, Count Toptani, introduced the saddle which goes with it. It has a deep, shortened seat with a high cantle, and the points come forward at an angle from the pommel, placing the stirrup about three inches further forward than on the hunting saddle. There are pads for the rider's knees and his weight is kept well forward. The tree is made with springs, giving a very comfortable seat. There are so many photographs of these saddles, used in show-jumping especially, that most people know well enough what they look like. The racing saddle is much flatter, but has the flaps cut right forward in the same way.

A lady hawking (from Newcastle's 'Horsemanship', 1667), showing the old-fashioned side-saddle with one pommel only, the right foot being crooked up to obtain some grip.

SIDE-SADDLES AND PILLION

Although the French marshal Villars fought the battle of Malplaquet on a side-saddle, because his wounds precluded him riding astride, and one African tribe also use the lance sitting sideways, this type of saddle has normally been designed for and used by women only. To a very great extent they were given up soon after 1900, although a number of women still ride hard to hounds on side-saddles.

The side-saddle is extremely comfortable, easy to learn on and gives a stronger seat than a man's saddle. The reason for this is that with a modern type the knees and thighs of both legs are exerting a very strong and comfortable grip on two wide pommels set fairly close together. The lower leg also has the support of a stirrup. It is an illusion that the seat on the horse is crooked on a 'lady's saddle'. The hips should, in fact, be as square across the horse as on any other type of riding saddle.

One theory of the origin of side-saddles is that, as soon as early packsaddles had some sort of framework to which to lash the loads, they also had some form of horn on each arch to facilitate the secure fixing of the load. One has only to visualise a woman leading a packhorse or mule to realise that she could easily hoist herself on to the pack and crook one leg round the horn of the front arch. The word pommel has for many years applied both to the arch itself and to the horn or crutch round which the woman put one leg, usually the right one. Before long women had rather better shaped packsaddles made for their sole use with the horn or pommel curved slightly to the off or right side. And the fact remains that well into the eighteenth century women are shewn holding horses with a flat seated saddle and one pommel as just described. A woman's mount, often a hinny—the offspring of a horse stallion and a donkey mare— was normally trained not only to walk fast but also to amble, the racking gait with the mount moving the two legs of the same side together. Trotting with one pommel would be disconcerting to say the least of it.

Presumably women would have started to ride on packsaddles by the end of the fourth century A.D. and about two hundred years later would have added a stirrup to the saddle, which would have helped a great deal to steady the left leg. Towards the end of the sixteenth century Catherine de Medici, queen of France, introduced a second pommel on the near side just below the top, or first pommel. The two pommels were now curved inwards towards each other so that the right leg, with a habit skirt and, in Victorian times at any rate, several petticoats wrapped round it, was fairly well wedged in between the two (see plate 6). Yet this innovation took a very long time to come into general use and was not common in Great Britain until the turn of the eighteenth century.

It was about 1830, the exact date is uncertain, that Mr. Fitzhardinge Oldacre in England invented a third pommel. It was placed below the second pommel on the near side with the curve downwards to fit the left leg, now brought up with a shorter stirrup and pressing against the new pommel. The second pommel just above it was still curved upwards to take the right leg and the pressure of the two legs towards each other gave the very strong seat already mentioned. The original first pommel no longer exercised any function and was gradually first reduced in size and then by 1900 in most cases abolished, though it persisted in other parts of the world for some time. (There are two other claimants for introducing the new pommel, M. J. C. Perrier and M. Baucher, both French experts.)

It remains to mention the habit of riding pillion behind a man, surely one of the most uncomfortable methods of travel ever invented. Strapped on behind the man's saddle was a small padded seat with a small rest for the feet and a handle at the back of the seat. The woman sat sideways holding on to her escort's belt with her right hand and to the handle, a rail, with the other. This contraption was in use from Tudor times, and probably much earlier, until well into the last century. Specimens covering this period are still in existence.

PACKSADDLES

Before the harness, possibly before the yoke, was the packsaddle for ox, donkey and in time nearly all animals tamed by man. Long before the riding saddle men and women had fastened their possession on to their beasts of burden. If men live in places where wheels can never penetrate and if dropping goods by parachute or helicopter is too expensive, there will the packsaddle be found. Not long ago in Lesotho a grand piano was packed, strung on poles, taken on a four-day journey over the mountains, and brought out again, intact.

The great days of the packsaddle were in the East, before the Portuguese in their caravels opened the sea routes round

R. D. Barrett-Lennard

1. 'Back-seat' driver in Morocco! This shows the correct seat on a donkey's back.

2. *Oxen at work on a farm near Orvieto in central Italy.*

3. *A pair of oxen in Spain in 1965 showing the heavy wooden yoke strapped to the horns with padding to prevent the horns being rubbed.*

4. A Bernese mountain dog harnessed to a small cart conveying milk churns.

5. A Mongolian archer leading a camel with the sawbuck type of packsaddle (thirteenth century).

6. The eighteenth century side-saddle with two pommels only. The right leg was wedged between them, the left leg supported by the stirrup alone.

7. *A knight in armour by Albrecht Durer (1498). The saddle is of the 'Newcastle' type. (Vienna, Albertina.)*

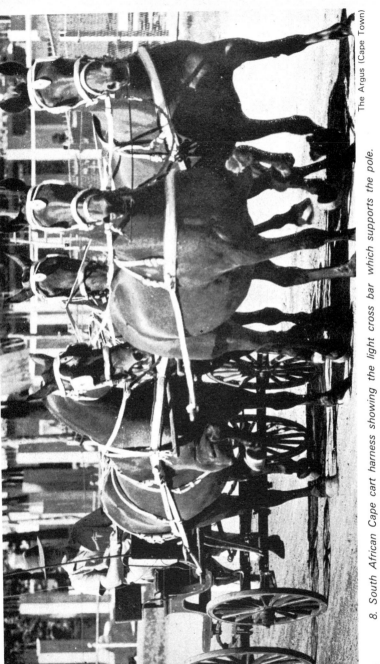

The Argus (Cape Town)

8. South African Cape cart harness showing the light cross bar which supports the pole.

9. *Present-day stage coach showing all the details of the harness.*

Raymond Lea

10. The head of an English hunter, showing a snaffle bit with a running martingale and separate nose-band.

Alan Loxley

Africa and took away from the Venetians the handling of the last stages of the immensely lucrative trade along the overland route of Serindia, the golden road to Samarkand, which brought the spices and treasures of the East to the Mediterranean ports for shipment to Europe via Venice.

When dealing with riding saddles it was possible to make a sharp division between those without trees and those made after the saddle tree was discovered in the fourth century A.D. Such distinctions are quite impossible with the packsaddle for there have always been and still are two schools of thought—those who prefer what we may call the primitive type with no tree in the real sense of the word and those who claim that the closer the packsaddle tree is to that of the riding saddle the better.

The difference between a riding saddle and a packsaddle is that the rider moves about constantly in the saddle, can get off and on at almost any time and can offsaddle for a short spell also at any time. The very opposite is the case with the pack animal. The weight carried is a dead one and should not be able to shift at all during a whole day's travel. So the greatest care must be exercised in fitting the saddle and in securing it and a sharp watch must be kept to see that the load remains always in the same place.

It is not possible to say when any particular form of treeless saddle was evolved for pack work, but various authorities consider that probably the best known is the Spanish *aparego*, *albarda* or *asparego*, still very much in use. It consists of two bundles, about four inches in diameter, of carefully selected long straw, without any cracked straws and with the heads cut off, each sewn up in a case of coarse linen long enough to reach from the withers to the hip. The bundles are joined at each end by a padded band. A large double pad of sacking stuffed with straw is placed over the bundles. Under this 'saddle', next to the horse or mule, goes a piece of linen and between the bundles and the sacking bags any blankets or cloths are placed. A surcingle comes next. Then the two biggest packages are roped on with a complicated tie known as the diamond hitch, leaving a hollow over the backbone. Any small packages are then added and also roped on, the rope also doing duty as a girth. Another version in use in America has a leather panel ribbed with willow wood and stuffed with hay. This takes the place of the bundles described above. This panel goes over a folded blanket and has girths, crupper and breeching of canvas.

Although the riding-saddle tree, with the side-bars rather

41

longer and the arches made with horns for convenience of lashing, is eminently suitable for packing there is another and possibly older form of packsaddle tree. This is the so-called 'sawbuck', used sometimes with the Spanish type *aparego* and thought to be of Arabian origin. The sawbuck is shown in a Chinese painting after a picture of the thirteenth century (plate 5). A Mongol horse archer, riding with a very short stirrup, is leading a Bactrian camel with the sawbuck pack-saddle most carefully delineated. It is in use today in many places, is light, weighing about seven pounds, easy to make and can be used with any type of pad. The sawbuck derives its name from the carpenter's sawhorse. It consists of two Xs, one at each end of the saddle, taking the place of the arches of the riding saddle. The top half of the X makes the pommel or cantle and is admirable for lashing the load on. The bottom halves of the Xs are joined by lateral bars which take the place of the side-bars but are not shaped and require heavy padding. The whole of this framework is made of light wood. In the painting mentioned above there are three Xs. The sawbuck is only suitable for light loads, but the camel saddle in use all over the East is similar with two horns back and front and with adjustable triangles on the animal's back.

Military packsaddles

The modern army packsaddle is a very different proposition; the British and American types owe a great deal to a saddle designed on the New Zealand goldfields in 1864, known as the Otago saddle. The original model had stirrups and like the Spanish *aparego* of today was frequently used as a riding saddle. The modern packsaddle is simply an adaptation of the cavalry riding saddle with steel trees jointed at the arches. Their great importance in the past was their use for trans-porting mountain artillery. This means heavy weights and special fittings. It is a fine sight today to see a Swiss army mountain battery using horses.

Packsaddles, either military or civilian, are fitted with different types of framework to take a stretcher each side or a chair facing forward for invalids. Probably of Turkish origin, they are called cacolets. In the eighteenth century there were saddles with fittings like large cacolets to seat a woman each side with her feet on a board facing forward with enough room for the very voluminous skirts. The horse or mule was led and the equipage took up a considerable amount of room. Before there were such things as mountain railways ladies were taken up such mountains as Mount Righi, each person sitting sideways on a packsaddle with very high pommel and cantle and a board or planchette for the feet.

It is fitting that this chapter should close with an account of the gear of the most impressive of all pack animals, the elephant. Although used in draught, the elephant is much better fitted for pack. Standing as much as eight feet high and able to carry up to 1200 pounds, of which about 140 pounds would be accounted for by the saddle, elephants are incredibly easy to load. Originally fitted with pad saddles in the 1880s, they were, in India on government service, issued with what is probably the simplest form of saddle tree ever used. Two pads or pillows, well stuffed and covered with waterproof material, are placed directly on each side of the spine and joined back and front by T irons, connected by side-bars, the whole frame being four feet long. The spine is entirely free and the bases of the two pads are so wide and the conformation of the ribs so suitable that with a load of 1200 pounds an elephant has marched several miles on level ground without girths, cruppers or breast straps. It is also possible to carry 200 pounds on one side and nothing on the other without the load shifting. Seats can be fitted to carry passengers. By 1955 in Indo-China the metal parts were made of aluminium with a mechanism which allowed for adjustments to be made in the case of animals of different sizes. In one of the official handbooks it is noted that the elephant saddle offers a perfect example of how a saddle should fit an animal.

BIT AND BRIDLE

A saddle need not be in the least necessary when using a horse in peace or war, and the same is true of bit and bridle. A horse can be, and countless numbers have in the past been, broken to obey the rider's will unequivocally with neither bridle on the head nor bit in the mouth with reins going back to the rider's hands. Voice, especially tones of voice, pressure of the knees and heels, shifts of balance and touches on the neck with the hand have all been used to guide and control horses long before saddles came in, and are still very much in evidence.

The bit goes in the mouth and generally speaking is of two kinds, a thick cord or a bar working on the corners of the mouth, known as a snaffle, or the more complicated curb bit, a bar or mouth-piece passing through the mouth and like the snaffle resting on the bars of the mouth above the teeth and on the tongue, with a chain or strap attached at each end

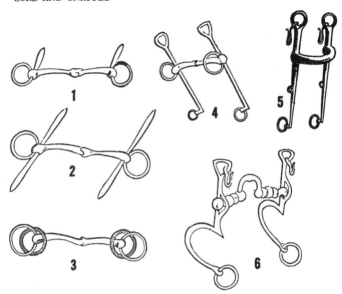

1. *Half-horn snaffle, best for the hunting bridle. 2. Plain snaffle. 3. Ring snaffle. 4. Plain-jointed Pelham. 5. Variation of curb bit. 6. Hanoverian Pelham. (S. Sidney, 'Book of the Horse).*

and resting in the chin groove (see illustration). The mouthpiece has straight pieces of metal, the cheeks, running at right angles downwards from each end and by applying force to their ends pressure is brought to bear on the curb chain in the chin groove. The mouthpiece of the curb bit often has a U-shaped arch in the centre, the port, which presses on the roof of the mouth and can be very severe. There are many variations and combinations of curb and snaffle, but the above are the two main types used for both riding and driving.

All bits have loops or rings, in the case of the curb bit at the top of the cheeks, fastening on to the headstall or head of the bridle, a strap going up the animal's head over the top behind the ears (see plate 10). This strap is in two parts, buckled together, the cheek piece and the crown piece which goes over the head. From the crown piece a band, the browfront or forehead-band, goes round the forehead above the eyes; and has loops for the crown piece to pass through. Also

44

running from the crown piece is a long strap, the throat lash, going under the jaw high up. Finally, a nose-band runs from the cheek piece round the head above the mouth. Apart from the action of the bit or bits the nose-band can give control by lowering its position and tightening it.

The cheek pieces, crown piece, throat lash, brow-band and nose-band constitute the headstall and can be used without the bit or bits as a halter, the animal being tied up by a rope passing through the nose-band and throat lash. With the bit and reins the headstall is the bridle. The reins are long thongs, or bands running from the bits to the hands of the rider or through rings on the harness to the driver in the vehicle or walking behind the team.

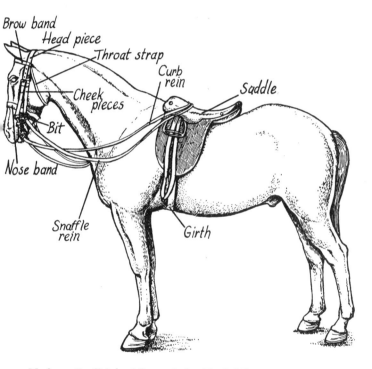

Modern English saddle and double bridle.

Once one has grasped that the bridle, the headstall and the halter are all more or less the same thing, one can, if one wishes, go into the various types of bit and methods of control used through the ages. There is also a form of halter consisting of cheek pieces with a crown piece running back at an angle behind the ears with a throat lash and nose-band but no brow-band. A short piece runs down from the throat lash to the nose-band. Cattle are often fitted with this simple pattern of halter or with the ordinary type, but a noose on the horns is more usual and in the case of a pack ox a hole is bored through the cartilage of the nose and a thin cord run through. In the case of a bull, and sometimes oxen, a metal ring is kept in the hole permanently.

Harness bridles

There is a further attachment to a harness bridle which was not, as far as is known, in use in ancient times. This is the blinkers or winkers, curved rectangles of leather fastened on the crown and cheek pieces projecting slightly outwards and designed to prevent the animal seeing behind. The idea is that a horse or mule is easily frightened by seeing as well as hearing a vehicle close behind from which there is no escape, except possibly by running away! This theory has been proved over and over again to be entirely fallacious, as witness the teams of the King's Troop, Royal Horse Artillery, in London today. But it persists and blinkers are still in use all over the world!

The most generally used bit is the snaffle, often made for driving with a solid, almost curved, mouthpiece and a ring on each end for the cheek pieces and reins. The most usual form of curb bit for harness is the pattern known as a Portsmouth, which is a curb bit with cheeks and curb chain and a ring at the top of the cheek which gives the same effect as a straight snaffle if the reins are buckled there.

Four-in-hand

A great deal of purchase on the bit is essential with one or more horses in harness and this is obtained by passing the reins through rings on, originally, the short neck yoke, and later on the upper part of the band of the breast harness where it goes over the shoulder, or on the collar harness and on the pad saddle. The rings on the pad saddle are called terrets, and there is one on each side; with a pair of horses the inside ones take the rein across to the other horse. With the four-in-hand the leaders have a separate pair of reins, going through the outside terrets of the wheeler's pad saddle and then through rings on the top of the throat lash of the wheeler's bridle on to rings on the collar of the leaders. These reins are split

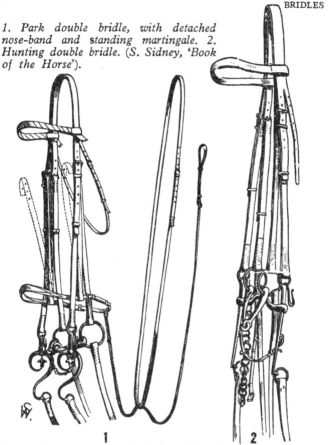

1. *Park double bridle, with detached nose-band and standing martingale.* 2. *Hunting double bridle. (S. Sidney, 'Book of the Horse').*

1 **2**

into pairs, as are the wheeler's and in the case of teams of more pairs the leaders' reins of a four-in-hand are continued on to the front pair, the leaders.

The nearer the front of a horse's head is to the perpendicular the easier it is to control him. In harness this is effected by running a bearing rein from the bit up to a ring on the crown piece and back to a hook on the centre of the pad saddle. These abominations can be so tightened as to give a horse very severe discomfort and they have been justly fulminated against for many generations. They were much in favour with

47

coachmen in the Victorian age who liked to 'hit and hold' their smart pairs of horses to make them step high. In state harness they were often passed through rings lower on the throat lash.

Except in certain cases riding horses do not have blinkers, though they are essential with some racehorses who do not like to see horses coming up behind them. The question of bits and bridles for flat racing is a very complicated one and is omitted from this book.

Double bridles

One of the most important differences between driving and riding bridles is that the latter are very often what are called double bridles or 'bits and bridoons'. The bridoon is simply a snaffle and the double bridle has both curb and snaffle in the mouth and two cheek pieces and headstalls and of course two pairs of reins. The Portsmouth bit gives much the same action as the double bridle and is very much used, especially by cavalry. The bit used by most of the world's cavalrymen has been the curb, which gives the most control to either the lancer or the swordsman. Men in armour always used a severe curb bit with long cheeks and not till the eighteenth century did the average cavalry soldier use any sort of snaffle. The exceptions are the Cossacks, descended from the Mongol invaders, who use the snaffle by itself. The single curb is popular all over the Americas, indeed in the Argentine it is used for racing on the flat. In Europe and in North America some form of snaffle is used on the Turf, but civilians and cavalry have a wide choice of double bridles or the Portsmouth type giving both forms of control. Although as far as is known the ancient world used some form of snaffle, for centuries the bit used in Asia and North Africa has been the single curb.

A variation of the snaffle which is fairly common is the gag, which has the same action as the bearing rein. The gag snaffle has a separate head with rounded cheek pieces, passed through projections on the rings of the bit and running thence as reins to the rider. Although valuable in the hands of an expert the gag can be terribly severe.

Two types of curb in use in the East are the Mameluke and the thorn bit. The former, in use in Mexico, Turkey and the East, has an iron ring instead of a curb chain which goes through the mouth as well. The port of the mouthpiece is as high as three inches. The thorn bit is even more brutal; on the mouthpiece is a metal thorn. The pain inflicted by these two bits can be imagined!

HORSESHOES

Although the metal horseshoe, in use in Europe since the fifth century A.D., is not part of the harness, it has played and continues to play such a large and important part in the use of horses, mules and to a lesser extent oxen, that it seems sensible to include an account of it in this book.

An arch-like frame of iron or other metal fitted to a horse's hoof as protection, the horseshoe has, apparently from a very early stage of its use, been endowed with various magic qualities. Its luck-bringing propensities are widely believed in, especially in the case of weddings, as a personal ornament, as a vital charm to be nailed (points uppermost so that the luck cannot spill out) over stable and other doors and even on the masts of sailing ships. Nor are there many countries where they do not obtain credence. Sometimes the points are shown turned down but the reverse seems to be the most popular. In some countries the horseshoe is considered to represent the crescent moon and, among other things, to have power to ward off the evil eye. A curious old fashioned belief current in England is that a fern, known as moonwort, has the property of pulling the shoes off any horse ridden over it.

There is much myth and magic about the metal horseshoe. The gods of the Norsemen were involved and their smith, Wayland, was credited with most useful powers. There are several of his so-called smithies in England; one is near White Horse Hill in Berkshire, where a man is supposed to be able to tie up his horse, place a coin on a stone, go away out of sight and on returning find the coin gone and the horse perfectly shod!

Before we leave the subject it may be of interest to realise that stories of horse thieves reversing the shoes on a horse before riding off so that his tracks will show going the reverse way are without foundation. The whole subject was summed up by ex-Farrier Sergeant A. Wheeley in *The Field* (11th May, 1961) as follows, 'It is possible to make a set of shoes hind side foremost and fit them' (in order to show tracks pointing in the reverse direction). 'Clips would go on each heel instead of on the toe and would hold for twenty miles. It would be impossible to remove a set of shoes and tack them on cold even under the best conditions. Horses' feet are neither round nor equilateral and if the near side shoes were nailed on the off side the nail holes would not serve and the problem would not be solved.' This description makes it clear that the narrow curved piece of metal, which forms a horseshoe, goes about threequarters of the way round underneath the hoof and leaves the heels uncovered. There are usually two clips on the heels and one on the toe, but there are many variations according to the work required of the horse. From five to eight nails are driven through the insensitive base of the hoof, an operation requiring great skill to avoid placing a nail too high into the sensitive part. A hoof must not be allowed to grow too long as this would interfere with the natural use of the foot. An iron horseshoe weighs roughly about a pound and lasts from 25 days to a month. In the case of racehorses the ordinary shoe is replaced by an aluminium plate for the actual race.

The hoof of an ox consists of two claws, has far less hard nail than the hoof of a horse and an unskilled smith can very easily permanently lame a beast. The last smith to shoe oxen in England shod a pair that pulled a waggon advertising somebody's beef cubes. He was the only craftsman to be found able to do the job and during the 1914-1918 war he had been concerned with shoeing oxen in Italy. He was also an expert maker of mule shoes.

Although horses always have functioned both in peace and war and still do so without being shod, this becomes difficult and eventually impracticable in mountainous country. Sometimes a horse is shod in front only.

For working under icy conditions horses must be 'roughed'. To give a grip nails with projecting heads are used and the heels of the shoes are turned down, and known as 'calkins', or the shoes can be tapped and specially made steel wedges or points screwed in without removing the shoes. The appalling losses in horseflesh experienced by the French in the retreat from Moscow in 1812 were largely due to the non-provision of the materials necessary for rough shoeing.

The early history of shoeing horses is very obscure. As far as is known both Alexander the Great, c.300 B.C., and Genghiz Khan, in the thirteenth century, tied rawhide 'boots' or 'sandals' on the horses' feet when negotiating mountain ranges. There is definite evidence in museums that the Romans used 'sandals' woven of cord like modern beach shoes and also 'boots' made of hippopotamus hide, in some cases strengthened with plates of iron, and fastened with ligatures. Nero and Poppaea are said to have adorned this type of shoe with jewels.

Even in 1809 a British officer suggested that cavalry should carry a pair of hemp shoes for use in an emergency. Nothing came of it, but it would be interesting to know where he got his ideas. Although such shoes would presumably last longer than the straw 'sandals' used by the Japanese and other Eastern peoples, their life would be a short one. As late as the end of the First World War the South West African Constabulary were obliged to use 'boots' made of giraffe hide to enable a patrol to get their horses back to civilisation from the northern part of the recently conquered territory.

It is perfectly possible for cavalry to operate with unshod mounts provided that there are sufficient remounts to replace those whose feet have given out. Genghiz Khan is said to have had as many as eighteen of the famous Mongolian ponies per trooper and the South African Republicans in the 1899-1901 war had up to five ponies apiece. As late as 1768, Cossacks in the Russian army rode unshod ponies.

When and where the technique of nailing metal shoes on to the hoof was first practised seems to be quite uncertain. Most authorities, if they make any statement on the subject, favour the Near East or the Mediterranean as the most probable locality. It has been stated that a horseshoe of modern type was dug up in England in the mid-seventeenth century in a grave reputed to be that of Childeric, who lived in the fifth century A.D. Whether this is correct or not it is more or less accepted that at this time such shoes were in use in Europe. Perhaps even more puzzling is the whole problem of quite a different form of iron shoe, still to be seen in the Near East in some places. This shoe is described and illustrated in great detail in Mayhew's *Horse Management,* though little seems to be known of the origin or even the approximate date of introduction. It is believed to have originated in Arabia and to have spread to the southern littoral of the Mediterranean, to Persia and to Portugal. The shoe consists of a piece of sheet iron stamped, not forged, covering the sole of the foot, with variously shaped orifices and attached by at least eight very thin nails driven through the hoof as in the case of the modern shoe, but going the whole way round the foot. Mayhew supposed that these plates were designed for the desert where there would be very little suction. They are all closed at the heels. Mayhew also shows an 'Old English' shoe used in the eighteenth century, not unlike the Arabian type but with calkins and a narrow opening at the heels and with holes for fourteen nails. It is reasonable to assume that specimens of this shoe, which is still extant, may be found in museums in some of the countries mentioned above. Whether the modern type of shoe came from the East it is impossible to say. In a history of the Byzantine Empire, published in 1854, there is a note that one of the

Seljouk Turkish armies which were threatening the Byzantine Empire in about A.D. 1000 should be attacked at once as 'the cavalry horses of the force were unshod'.

Although horseshoes were common in France and England in the eleventh century—the cavalry horses in the Bayeux Tapestry are all shod—horseshoes did not reach Scandinavia till the next century. They used frostnails in eastern Scandinavia before shoes were in common use.

It is obvious that metalled roads of any sort need shod animals to work on them, and the more roads there were the more work there would be for the farriers. A pack animal can on many stretches be led on the comparatively soft verge of a metalled road, but wheeled vehicles must keep on the metal and the teams that draw them must be shod if they are to last for any length of time.

A by-product of the farrier's highly skilled trade in the eighteenth and early nineteenth centuries was the importance of used horseshoe nails in the manufacture of the barrels of pistols of high grades. These nails, collected by children along the roads where traffic was heavy, were eagerly bought by the gunsmiths and welded into the required shape.

BIBLIOGRAPHY

The Age of Elegance, by A. Bryant; Collins, 1950. For the old coaching days.

Armour and Weapons, by ffoulkes; Oxford University Press, 1909.

Book of the Horse, by Sidney; Cassell, 1878. Once a standard work and very useful.

Britain in the Roman Conquest, by Liversedge, 1968.

Country Blacksmith, by Niall, illustrations by Meirion Roberts; Heinemann, 1966. Especially for shoeing oxen.

Fighting Men, by Treece; Brockhampton, 1963.

First Book of Irish Myths, by Eoin Neeson; Mercia Press, Cork.

Handbook for Military Artificers, H.M.S.O., 1915.

History of Mankind, Vol. 1, part two; Allen and Unwin, 1962, under the auspices of Unesco. Invaluable for the very early periods when harnesses were being 'discovered'.

History of Tourism, by Sigaux; Leisure Arts, c.1966.

Horseman's Dictionary, by Bloodgood and Santini; Pelham Books, 1963.

Horses and Saddlery, by G. Tylden; J. A. Allen & Co., 1965. Deals with all British army saddlery and harness.

Horses of the World, by Goodall; Country Life, 1965. The illustrations cover the whole range of modern horse harness in great detail.

Horseshoeing as it is and should be, by Douglas; Murray, 1873.

Illustrated Horse Management, by Mayhew; 17th edition, Glaisher, 1901. Especially for horseshoes.

Just Elephants, by W. Baze; Elek Books, 1955.

L'Ecole de Cavallerie, by de la Guériniere, 1733; see under *White Stallions of Vienna*.

Legion of Frontiersmen's Pocket Book, by Pocock; Murray, 1909.

Manual of Horsemastership; H.M.S.O., 1937.

Mediaeval Costume, Armour and Weapons, 1350 to 1450, translated by Jean Layton from the German; P. Hamlyn, 1958.

Military Opinions, by Bourgoyne; London, 1859. For details of Spanish locally-made packsaddles.

More About the Forward Seat, by Littauer; Hurst and Blackett, 1947.

Notes on Transport and Camel Corps, by Burn; H.M.S.O., 1887.

Ordnance—The History of Army Ordnance Services, by Forbes; Medici Society, 1929.

Philipson on Harness, Nimshivitch on the Cape Cart; published Stanford, London, 1882. Especially for coaching in England.

Points of the Horse, by Hayes; new edition, Stanley Paul, 1969.

Riding and Hunting, by Hayes; being republished. Hayes' two books are standard works.

'The Silver Hand of Alexander', by Bethell, *Blackwood's Magazine*, May 1928. Republished in *His Majesty's Shirtsleeves*. Contains an account of the old Asian trade routes.

White Stallions of Vienna, by Podhajsky; translated Hogarth-Gaule; Harrap, 1963. Especially for the extracts dealing with L'Ecole de Cavallerie, see above.

INDEX

54

ACKNOWLEDGEMENTS

The author wishes to acknowledge the very material help received over illustrations and books of reference from his wife, from Mrs. G. D. Morgan and Mrs. S. Michaelson, his daughters, and from Kate Bergamar, author of *Discovering Hill Figures*. He is also much indebted to the last named for introducing him to the legend of the fern moonwort, and to Mrs. J. Vaughan for the myth of the ox yoke in Ireland.

Printed by C. I. Thomas & Sons (Haverfordwest) Ltd., Press Buildings, Merlin's Bridge, Haverfordwest, Pembrokeshire.